Garfield's Almost-as-Great-as-Naps Guide to ENGINEERING

Rebecca E. Hirsch

Garfield created by JIM DAVIS

LERNER PUBLICATIONS ◆ MINNEAPOLIS

In memory of my mother-in-law, Alice Hirsch

Copyright © 2020 Paws, Incorporated. All Rights Reserved. "GARFIELD" and the GARFIELD Characters are trademarks of Paws, Incorporated.

Visit Garfield online at https://www.garfield.com

All rights reserved. International copyright secured. No part of this book may be reproduced, stored in a retrieval system, or transmitted in any form or by any means—electronic, mechanical, photocopying, recording, or otherwise—without the prior written permission of Lerner Publishing Group, Inc., except for the inclusion of brief quotations in an acknowledged review.

Lerner Publications Company
A division of Lerner Publishing Group, Inc.
241 First Avenue North
Minneapolis, MN 55401 USA

For reading levels and more information, look up this title at www.lernerbooks.com.

Main body text set in Neo Sand Std 13/20.
Typeface provided by Monotype Typography.

Library of Congress Cataloging-in-Publication Data

Names: Hirsch, Rebecca E., author.
Title: Garfield's almost-as-great-as-naps guide to engineering / by Rebecca E. Hirsch.
Description: Minneapolis : Lerner Publications, [2020] | Series: Garfield's fat cat guide to STEM breakthroughs | Audience: Ages 7-11. | Audience: Grades 4-6. | Includes bibliographical references and index.
Identifiers: LCCN 2018047657 (print) | LCCN 2018058842 (ebook) | ISBN 9781541561892 (eb pdf) | ISBN 9781541546400 (lb : alk. paper) | ISBN 9781541574281 (pb : alk. paper)
Subjects: LCSH: Engineering—Juvenile literature.
Classification: LCC TA149 (ebook) | LCC TA149 .H5 2020 (print) | DDC 620—dc23

LC record available at https://lccn.loc.gov/2018047657

Manufactured in the United States of America
1-45569-41278-1/14/2019

Contents

Great Pyramid 4

Roman Engineering 6

Great Wall of China 8

Machu Picchu 10

Steam Power 12

Electric Power 14

The Assembly Line 16

Panama Canal 18

The Internet 20

Channel Tunnel 22

International Space Station 24

Large Hadron Collider 26

Breakthrough of the Future? 28

Paws-On Project 29
Glossary 30
Further Information 31
Index 32

Great Pyramid

About four thousand years ago, ancient Egyptians designed and built the Great Pyramid at Giza. Workers harvested 2.3 million stone blocks. They brought the blocks to the desert on boats. The boats traveled along the Nile River and a series of canals. On land, workers pulled the blocks on heavy sleds. They wet the desert sand to make the sleds easier to pull. Then they raised up the blocks on ramps as the pyramid grew taller.

ANCIENT EGYPTIANS WORSHIPPED CATS— OBVIOUSLY AN ADVANCED CIVILIZATION!

Building a pyramid before the era of modern construction was no easy task!

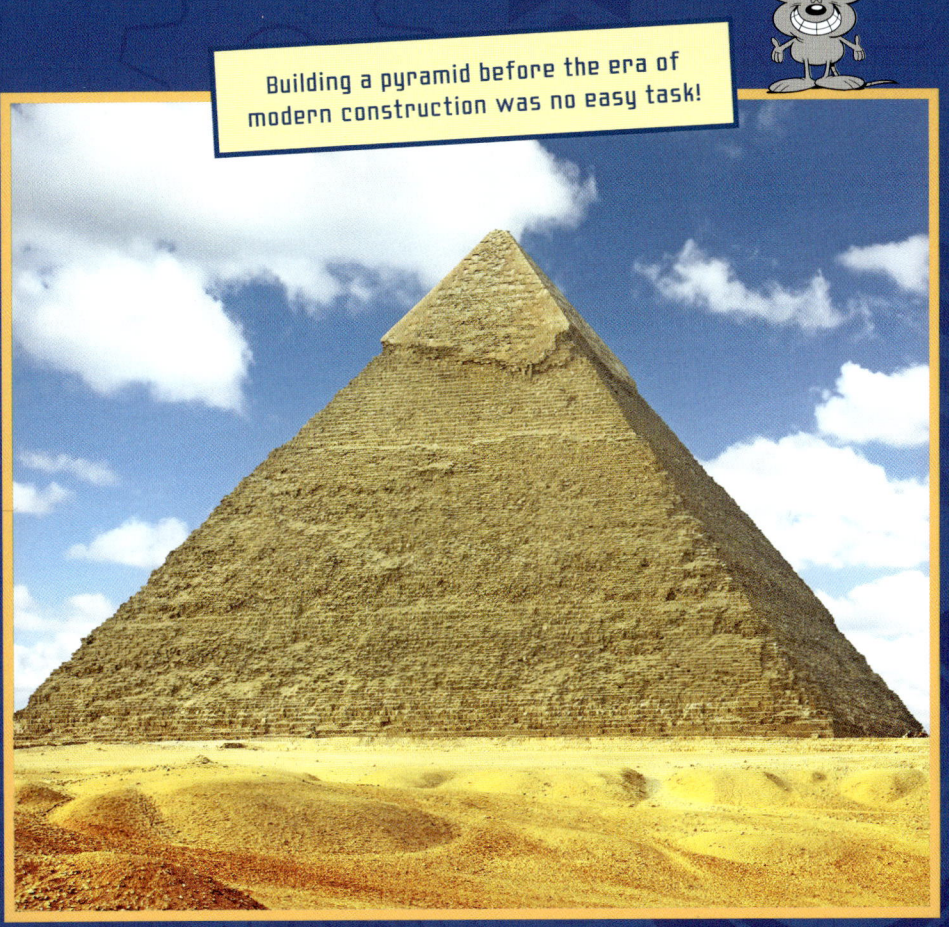

When the pyramid was finished, it stood 481 feet (147 m) and weighed 5.8 million tons (5.2 million t). It remained the tallest building in the world for more than thirty-eight hundred years.

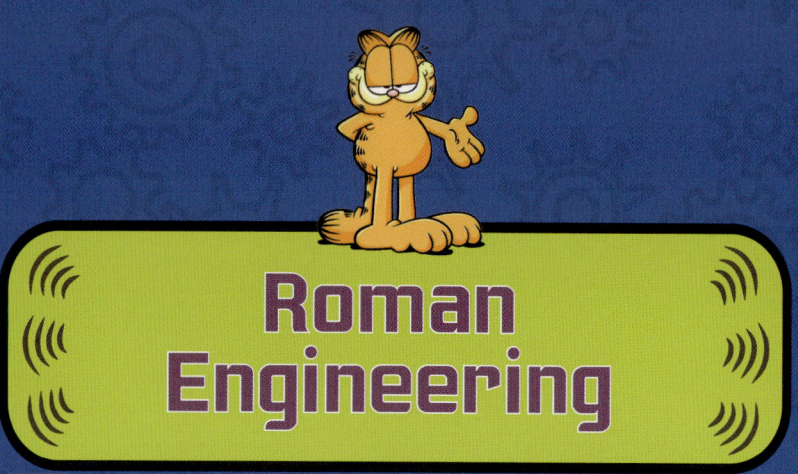

Roman Engineering

More than two thousand years ago, the ancient Romans performed great feats of engineering. They built aqueducts, or channels for bringing water to cities and towns. The water was used for public fountains and baths. A detailed system of tunnels and pipes moved the water by gravity. Some aqueducts crossed over valleys on high stone arches. Some of these arched aqueducts are still in use.

DID ANCIENT ROMANS HAVE RUBBER DUCKIES TOO?

Roman construction is an example of early engineering at its best.

The Romans also built a network of roads. The network stretched from southern Italy to Britain and from the Middle East to Spain. More roads ran across northern Africa. The Romans built 50,000 miles (80,470 km) of roads.

Great Wall of China

The Great Wall of China is the world's longest human-made object. The wall was built across northern China to keep out invaders. A massive workforce of soldiers, prisoners, and local workers spent more than two thousand years building the wall out of bricks and packed earth.

The Great Wall is not actually a single wall but a series of walls. Some of the walls overlap, running parallel to one another. When surveyors measured all the walls, they discovered the Great Wall is 13,173 miles (21,200 km) long—more than twice as long as people had thought it was!

Achievements in IGNORANCE

The Leaning Tower of Pisa created headaches for Italian engineers from the beginning. The tower began to lean in 1178 when only the bottom three floors were complete. The ground under it was too soft. The engineer in charge tried to fix the problem. He made the upper floors slightly taller on the short side. But that just made one side heavier, and the tower leaned even more. Hey, even engineers make mistakes!

> I'M GLAD THERE'S NO LEANING TOWER OF ODIE... THE GROUND THERE WOULD PROBABLY BE TOO SOFT FROM ALL THE SLOBBER!

These days, the Leaning Tower of Pisa is one of the world's most well-known buildings.

Machu Picchu

In the fifteenth century, the Inca built Machu Picchu, a city in Peru high in the Andes Mountains. It may have been a royal estate. At some point, the Inca abandoned Machu Picchu. An explorer discovered the lost city more than a hundred years ago.

BUILDING MACHU PICCHU SOUNDS LIKE A LOT OF WORK... I'M ALLERGIC TO WORK

Can you imagine hauling stones up a mountain to build a whole city?

Machu Picchu contains about two hundred buildings. More than three thousand steps run through the city. The city was built without metal tools or the wheel. Large stones were transported up the mountain. The stones were cut and fit together without mortar. The stones fit so tightly together that a knife blade can't fit between them.

Thanks to its excellent stonework, Machu Picchu can stand through earthquakes, which are common in Peru. Without the high-quality stonework, the buildings would have collapsed by now.

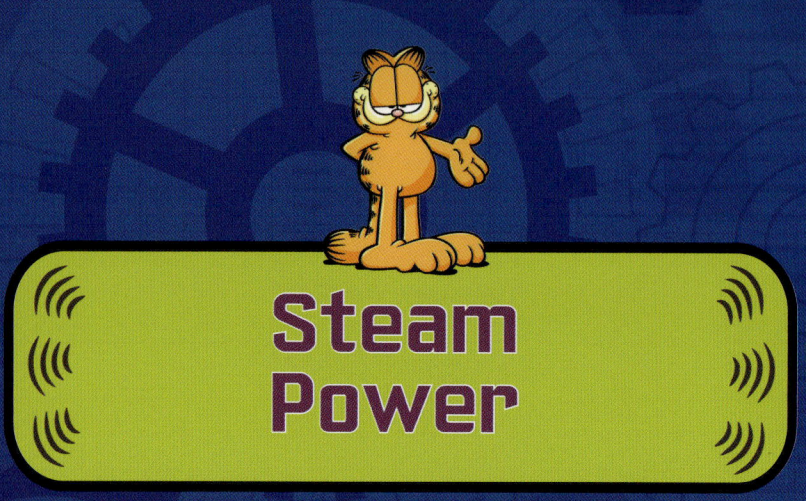

Steam Power

For a long time, people had realized that steam contains energy. Some thought steam could even be used to do work.

In the seventeenth century, people began to harness the power of steam. In 1698 English engineer Thomas Savery built the first working steam engine. His engine burned coal to make steam and used steam power to pump water out of flooded coal mines.

I PREFER CAFFEINE POWER!

Steam locomotives were an engineering feat of their day.

In the eighteenth century, steam engines powered factories, ships, and locomotives. By the early twentieth century, steam-powered automobiles even ran on roads. These cars were so fast they set speed records that lasted for a century!

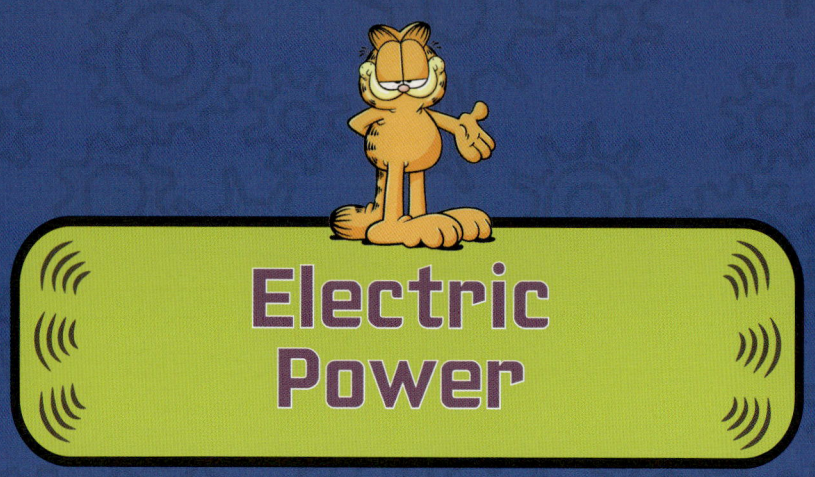

Electric Power

In 1752 Benjamin Franklin tied a key onto a kite string during a storm. He proved that lightning and electricity were the same thing. But could people make electricity and put it to work?

In 1831 English scientist Michael Faraday discovered he could make electricity by passing a magnet through a copper wire. In 1882 Thomas Edison opened the world's first electric power plant in New York. It could power five thousand lights.

The use of electric power continued to spread. Now, with the flick of a finger, people can tap the power of electricity in homes and schools.

The Assembly Line

In 1913 not many people could afford automobiles. Henry Ford wanted to make his Model T more affordable. He broke down the assembly of cars into steps. Each worker was trained to do just one of the steps. A moving belt carried the parts. Workers assembled the Model T one step at a time as the cars passed down the assembly line.

I WANT A LASAGNA ASSEMBLY LINE, AND THE LAST STOP IS MY MOUTH

Model Ts don't look much like modern cars, but they helped introduce the era of modern transportation.

Thanks to the assembly line, the Model T became quick to make and inexpensive. Ford's cars became affordable to more people, including Ford's own workers. Within about five years, more than half the cars sold in America were Model Ts.

Panama Canal

In 1914 the first ship sailed through the Panama Canal. The 48-mile (77 km) waterway created a shortcut between the Atlantic and Pacific Oceans. Before the canal was built, ships had to go all the way around South America. The canal cut about 9,200 miles (14,800 km) off the trip.

The Panama Canal

Building the canal was difficult. Workers used steam-powered machines to dig the canal and build huge concrete locks to raise and lower ships.

The Panama Canal remains a busy shipping route. More than thirteen thousand ships pass through each year.

The Internet

In the mid-twentieth century, there was no easy way to move information from one computer to another. Scientists and engineers began experimenting with ways for computers to talk to one another. By the 1980s, they were using computers to send files back and forth.

In 1991 a Swiss programmer named Tim Berners-Lee invented the World Wide Web. This was more than just a way to send files. It was a web of information that anyone could access. Today the World Wide Web brings a large amount of knowledge to people all over the world.

 NOW IF SOMEONE COULD INVENT A WAY TO BRING A LARGE AMOUNT OF DOUGHNUTS TO MY FINGERTIPS

Schools and libraries are great places for accessing the internet.

Channel Tunnel

In 1994 the Channel Tunnel opened between France and England. It is the longest undersea tunnel in the world. The underwater portion is 24 miles (38 km).

The Channel Tunnel

WHO NEEDS TRAIN TRAVEL WHEN THERE'S DOG TRAVEL?

More than thirteen thousand engineers and workers helped build the concrete tunnel. Machines drilled through rock in the seabed. The machines started on each side and met in the middle. One of the machines was too big to bring out. So it made a right turn, drilled away, and was left behind. The machine was cemented in!

The Channel Tunnel is actually three tunnels. Two are used for high-speed trains. A third is an emergency escape.

International Space Station

The International Space Station is a spacecraft in orbit around Earth. It is the largest human-made object ever put into space. The space station weighs almost 1 million pounds (453,592 kg). It is big enough to cover a football field!

TEAMWORK WORKS!

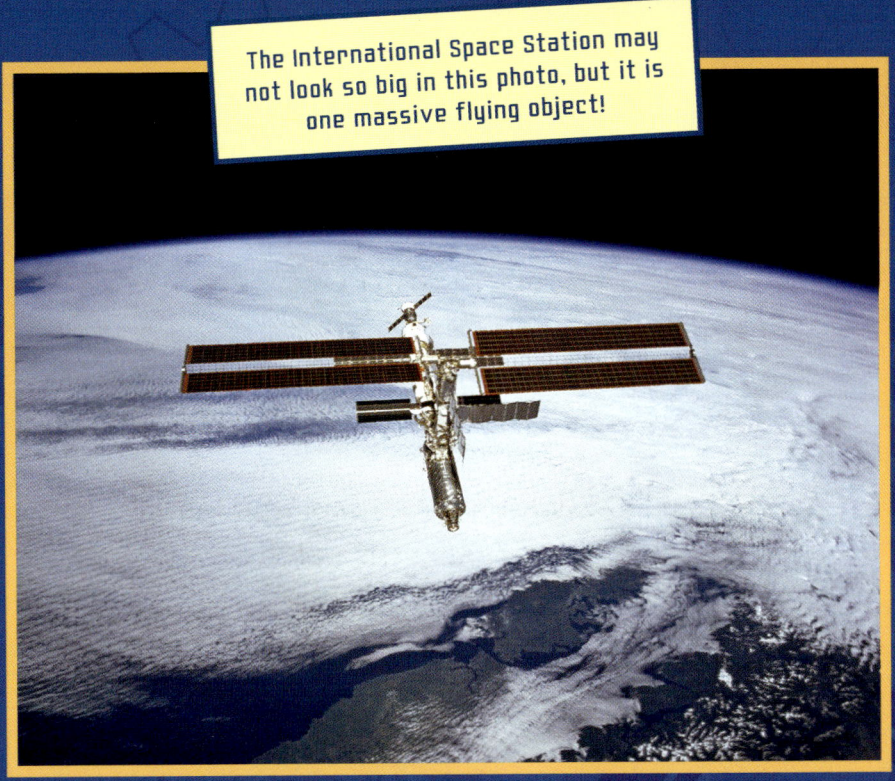

The International Space Station may not look so big in this photo, but it is one massive flying object!

Many countries worked together to build the station. The first part was launched in 1998. Then more parts were added. All the parts were put together in space.

The first crew arrived at the space station in 2000. People have lived on the station ever since. Scientists use the station to study how to live and work in space.

Large Hadron Collider

The Large Hadron Collider is the largest machine in the world. Completed in 2008, it is a circular tunnel 17 miles (27 km) long. It sits about 328 feet (100 m) underground between Switzerland and France.

I DON'T SMASH PARTICLES— JUST SPIDERS

The Large Hadron Collider helps scientists study the universe.

The machine uses powerful magnets to hurl the world's tiniest particles around a track. The particles reach almost the speed of light. Then scientists smash the particles together. They study the collisions to learn how the universe began and what it's made of.

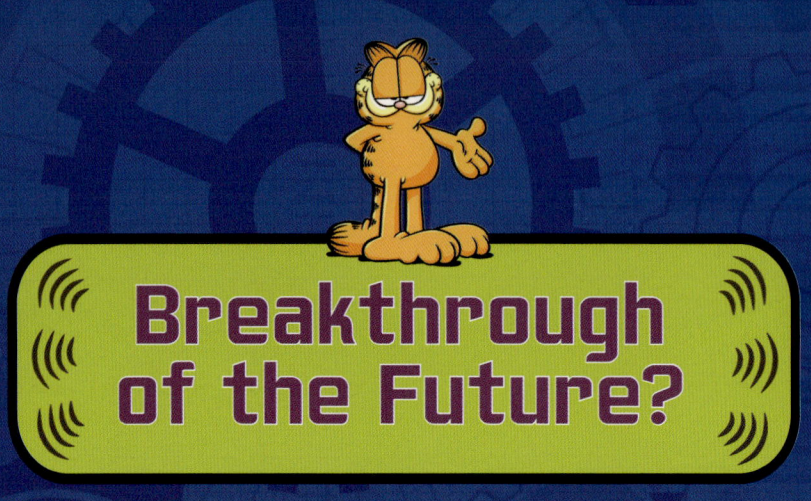

Breakthrough of the Future?

Imagine taking a dream vacation into space. Start planning now! Space vacations are already happening. The first tourist in space was an American businessman named Dennis Tito. In 2001 he flew aboard a Russian rocket and arrived at the International Space Station. But the cost of a space vacation remains out of this world. Tickets for a ten-day trip cost about $55 million. The cost will probably come down in the future, though.

PAWS-ON PROJECT

WANT TO ENGINEER A MODEL OF AN EGYPTIAN PYRAMID? The ancient Egyptians made theirs from stone blocks. For your pyramid, try sugar blocks. Start with a sturdy paper plate or a piece of cardboard as a base. Add a bottom layer of sugar cubes. Glue the next layer of blocks on with tacky glue or a hot glue gun (with an adult's help). Don't use regular school glue, or the sugar will melt. Keep adding layers of blocks until you reach the top!

Glossary

aqueduct: a structure that carries flowing water across a river or valley

canal: a human-made waterway for boats

engineer: a person who uses science and mathematics to design solutions to problems. Engineers design and build roads, bridges, or machines and create new products.

engineering: the science of using nature's power and resources in ways that are useful to people

lock: an enclosure with gates at each end, used for raising or lowering boats as they pass through a canal

locomotive: a train engine that can move under its own power and hauls railroad cars

parallel: lying in the same direction and always the same distance apart

particle: a very small part of matter

programmer: a person who writes computer programs

surveyor: a person whose job is to find out the size, shape, and position of objects on land

Further Information

Connolly, Sean. *The Book of Massively Epic Engineering Disasters.* New York: Workman, 2017.

Macaulay, David. *How Machines Work: Zoo Break!* London: Dorling Kindersley, 2015.

Roman Aqueducts: The Dawn of Plumbing
https://www.kidsdiscover.com/quick-reads/roman-aqueducts-dawn-plumbing/

Seven Wonders of the Ancient World
https://kids.nationalgeographic.com/explore/history/seven-wonders/#Pyramids-at-Giza.png

Turner, Matt. *Genius Engineering Inventions: From the Plow to 3D Printing.* Minneapolis: Hungry Tomato, 2018.

What Is the International Space Station?
https://www.nasa.gov/audience/forstudents/k-4/stories/nasa-knows/what-is-the-iss-k4.html

Index

ancient Egyptians, 4, 29
ancient Romans, 6, 7
Berners-Lee, Tim, 20
cars, 13, 16-17
computers, 20
Edison, Thomas, 14
electricity, 14
Faraday, Michael, 14
Ford, Henry, 16, 17
Franklin, Benjamin, 14

Inca, 10
Leaning Tower of Pisa, 9
roads, 7, 13
Savery, Thomas, 12
ships, 13, 18-19
space, 24-25, 28
steam power, 12, 13, 19
trains, 23
water, 6, 12
workers, 4, 8, 16-17, 19, 23

Photo Acknowledgments

Image credits: Savaryn/Getty Images, pp. 2, 3, 6, 7, 10, 11, 14, 15, 22, 23, 26, 27, 30, 31, 32; Zally23/Shutterstock.com, pp. 4, 5, 8, 12, 13, 16, 17, 18, 19, 20, 21, 24, 25, 28; Dudarev Mikhail/Shutterstock.com, p. 5; cge2010/Shutterstock.com, p. 7; QQ7/Getty Images, p. 9; Marco Straer/EyeEm/Getty Images, p. 11; mrtom-uk/Getty Images, p. 13; Everett Historical/Shutterstock.com, p. 15; catscandotcom/Getty Images, p. 17; Marian Stoev/EyeEm/Getty Images, p. 19; Rawpixel.com/Shutterstock.com, p. 21; EQRoy/Shutterstock.com, p. 22; NASA/Rodney Grubbs, p. 25; © Pascal Boegli/Getty Images, p. 27.

Cover Image: Savaryn/Getty Images.